John 10:10: A thief comes only to steal and to kill and to destroy. I have come so that they may have life and have it in abundance. (HCS)

HARVEST OF HEALING, LLC

Izauh 61™

Publishing assistance by BookCrafters, Parker, Colorado.
www.bookcrafters.net

In an Attempt to Save the World

LIVING BY THE LIGHT OF THE MOON

HARVEST OF HEALING, LLC

Izauh 61™

INDEX OF CONTENTS

INTRODUCTION

I Corinthians 3:19-20: For the wisdom of this world is foolishness with God, since it is written: He catches the wise in their craftiness; and again, The Lord knows that the reasonings of the wise are meaningless. (HCS)

The years 2024 and 2025 are earmarked for the next level of change in the world. Turbulent shifts in the atmosphere have been happening at a gradual pace since 2012, but bigger things will unfold at a more rapid rate soon, with a promise of a brighter future ahead. Like birthing pains, not always pleasant and sometimes seem intolerable but at the conclusion of the distress a new life will begin. Living in a fashion that mirrors ancient times and practices will begin to emerge as people come to the knowledge of the benefits the practices can provide. The stage is being set for rescue.

I Corinthians 15:51-52: Listen, I am telling you a mystery: We will not all fall asleep, but we will all be changed, in a moment, in the twinkling of an eye, at the last trumpet. For the trumpet will sound and the dead (speaks of Soul death, not physical death) *will be raised incorruptible, and we will be changed.* (CSB) (description added)

Exodus 3:17: And I have promised you that I will bring you up from the misery of Egypt to the land of the Canaanites, Hittites, Amorites, Perizzites, Hivites, and Jebusites – a land flowing with milk and honey. (HCS)

Egypt is the name used for the world systems we know today. "ite" stands for resident. This verse foretells of people being brought to a way of life that results in sufficient, beneficial supply.

Once Upon a Time

The past 12 years have brought many enlightenments through thorough studies and personal experiences with a focus on health. As I step back to scan the broad picture, my conclusion is that humans have evolved themselves into a species that experiences untimely decay and physical hardships. Once upon a time, however long ago that may have been, humans would have been vessels that contained vibrant, healthy cells. It appears that through convenient world systems, as warm and fuzzy as they may all seem, humans have advanced themselves into a species that requires more food, water, prescription drugs, sex, accumulations, socialization, and so forth. The recipe for life has become so overloaded with preservatives, meaning things people have come to believe are necessary for life to continue, the vitality and cycle of new cell growth within the body has become drowned out. The disturbing thing about this is that at the end of the live it up, live for the world and all that it provides lifestyle, one can meet their eternal end and no eternal life is left to be had.

Some may not have a concern about what happens upon their death, and that's okay. Where understanding is lacking is a few hundred years ago people may have been able to say a prayer prior to their earthly departure and make the trip to the great hereafter. Today, with the level of contamination that has afflicted the blood, I seriously doubt the trip beyond earth can be made and certainly not with just a prayer. Once upon a time people had healthier DNA strands than most of the population has today.

Making some adjustments to the current day lifestyle must take place to evolve the human species into a position that grants longer lifespan and healthy new cell growth. With this in mind, I gathered information from ancient Scripture and a picture developed that made sense. To avoid a breakdown in the mechanics, the Owner's Manual (Scripture) must be followed, isn't that how it is? Ancient practices must be resurrected and followed for the human DNA to become disease free and the body to have a healthy lifespan.

There are two terms used in Scripture that identify two species of humans that are now in existence. One species is rare at this point, with the other being the dominating pack. The title of Hebrew paints a picture of a species of human that has a physical body that seems to function more on elemental forces than on food and drink, and runs at an optimal level. Then there is the title of Samaritan that paints a picture of a group that is more flesh driven vs. elemental power driven. They require water and possibly a more frequent or greater quantity of food, and sleep patterns can be different than that of a Hebrew. This compares to vehicles that are powered by electricity vs. those powered by gasoline. Now the task is describing how to change the gasoline powered body for the electrically powered body.

Remodeling

What are the two most common world systems shared equally around the globe? The practice of 1) healthcare; and 2) religion. Not all forms of preferred healthcare are the same and certainly not all styles of religious practice are the same. The common goal is to be spiritually and physically healthy.

Given these two common systems and the frequency in which they are consulted, it makes sense that something amongst these two common goals has gone astray when considering the condition of health the majority of the population is experiencing. People around the globe suffer from some form of physical illness or disease at varying degrees. I thought spiritual health and physical health were found within the walls of religious institutions and healthcare facilities. Right? We are witnessing something contrary to what has been presented as the correct avenue to secure a healthy spirit, a Soul filled with light, and a healthy body. It appears there is a need for remodeling within these two world systems. And, like many other remodeling projects, you think the job will only involve remodel of the kitchen and you find behind the walls of the kitchen there are problems with the plumbing and the electricity. The projects required for the remodel extend throughout the house before the project is complete. In addition, the targeted budget for the remodel project is exceeded before all is complete. A demand for additional funds arises and the date of completion is pushed out a few more months. This is how the reconfiguration and remodeling of spiritual and physical healthcare will happen. A little at a time and the project will be much more extensive than what is originally anticipated. So where do we begin?

The following chapters will interpret many common terms used in Scripture but shed a new light upon their meaning. Star Dust is the term used to describe particles from the Heavens that become present in the blood. Let's begin.

CHAPTER I

DOING THE TANGO

It can be a challenging feat to unravel parables and prophesies that were written thousands of years ago when many of the terms we use today did not exist at the time those texts were penned. In lieu of written communication, verbal stories were told in a manner to shed light on the seemingly unexplainable situations encountered or witnessed. The ability to skillfully Tiptoe Through the Tulips of word play becomes essential and a map to connect the dots can become a tangled mess.

Deuteronomy 6:6-7: These words that I am giving you today are to be in your heart. Repeat them to your children. Talk about them when you sit in your house and when you walk along the road, when you lie down and when you get up. (HCS)

I will use a short parable at times to attempt a comprehensive explanation for what is presented. Many parables and prophesies link together creating an unending chain of events. To interpret it all in one book would require a text more voluminous than what the average person would have an interest in reading and more organization than one person can master. This book will focus on some of the more important details with the remainder of the chain of events for a later date. First things first.

CHAPTER II
MARCH TOWARD DEATH

Hosea 4:6: My people are destroyed for lack of knowledge (intelligence). *Because you have rejected knowledge, I will reject you from serving as My priest. Since you have forgotten the law of your God, I will also forget your sons.* (HCS) (description added)

Hosea is telling us that if we want to avoid being destroyed, we better figure out what is going on, learn it and apply it. With escalating numbers of untimely deaths, people obviously lack the knowledge Hosea is referring to.

Human health has been declining rapidly, coupled with the continuous rise of healthcare costs, including health insurance and pharmaceuticals, to a level many cannot afford. There is a never-ending cycle of healthcare that does not hold all the answers and does its share of contributing to death.

According to pubmed.ncbi.nlm.nih.gov: "Recent studies of medical errors have estimated errors may account for as many as 251,000 deaths annually in the United States, making medical errors the third leading cause of death." Considering the total number of people in the United States this number may not seem troubling, unless one of those numbers was your family member.

Cdc.gov reports a provisional total of 3,090,582 deaths in the United States in 2023 with heart attack and cancer in the number one and number two positions for cause of death.

In 2014/2015 a brother of mine was in the hospital. During his hospital stay the physician informed my brother that he thought it necessary to perform a colonoscopy to determine the cause of intestinal region discomfort. My brother, putting his trust in the physician, agreed to the procedure. During the procedure my brother told the physician that he was experiencing extreme pain. The procedure was completed, and the report was that nothing prominent was visible. The physician, not giving the painful experience much consideration, discharged my brother from the hospital. Within a few hours, my brother went to the hospital emergency room in extreme discomfort. What was later discovered was that during the colonoscopy the physician had punctured the wall of the intestine, causing toxins to seep into my brother's abdominal cavity. Emergency surgery was required. My brother had previously completed a season of lung cancer treatment, resulting in the wound from the abdominal surgery being challenged to heal. At one point the surgical wound had reopened resulting in the need for a second surgery to his abdomen. With an added hurdle of now having a colostomy bag, his health rapidly declined and by early 2016 the cancer had returned. My brother passed away in July of 2016.

This is just one example of the catastrophes that can take place inside medical facilities. Are there times when medical intervention is necessary? Absolutely. The point we must strive for is reducing, and hopefully, at some point, eliminating the invasive treatments that can result in catastrophes. An area lacking in knowledge is the level of impact invasive tests and high resonance imaging can have on the electrical activity (aura) within and around the body. The processes used to obtain an answer to a health problem may in fact be creating other, at times more serious, problems that likely will not become evident until months later. This delay in reaction swings the door wide open for denial by the medical providers of any counterintuitive response being a result from the tests conducted months prior.

What about religion? Numerous promises circulate amongst religious circles about qualifying for a great hereafter if you march to the beat of the drum of any given number of religions. The number of diseased and dying people have not declined inside these religious circles. What is happening here? False hopes? Or, the way to apply a beneficial spiritual health practice has been lost? When things, activities included, go counterclockwise with nature, eventually catastrophes will result.

CHAPTER III
TIME IS UP

How long will it take before people escape from the stronghold of modern-day healthcare and religion that is glaring them in the face? Having studied parables and prophesies found in the Holy Bible, along with glancing through myths (admittedly no expert in myths) coupled with the condition the world is in, the End of Days must be what is unfolding. People are literally being eaten alive, whether by financial stress, physical stress or relationship stress, it all wears on the physical and emotional body. If this End of whatever all that encompasses is not upon us and the birth of change is not soon coming, the entire collective group of humans and possibly the earth herself will likely fade away. Therefore, age old mysteries that have been highlighted by revelation are shared in an effort to kick start a New Beginning and get the momentum of life moving in a positive direction. Many hours of study have accompanied the mysteries discovered thus far, along with practices I have merged into my daily life. Someone must be the guinea pig, right?

CHAPTER IV
THE BLOOD CODE OF JESUS

There is a uniqueness in the blood of certain individuals. Whether it lies within the blood cells, blood plasma or both is yet to be determined. This uniqueness has what I only know to describe as an electrical charge that exceeds the boundaries set in place by life on earth, medical discoveries or diagnosis. I call this unique quality "Star Dust", a fragment of an element that comes from the Heavens (Cosmos); a force or energy that is unable to be measured or detected by any means of this current day. Something involving this element from Heaven took place during conception for some and results in an individual with different or unique qualities, sensitivities or reactions to various things in the environment. For others, a gradual uptake of this element is at play when conditions are conducive.

Jeremiah 1:5: I chose you before I formed you in the womb; I set you apart before you were born. I appointed you a prophet to the nations. (HCS).

Moon cycles, planet alignment and activity beyond description initiate all types of events that humans have identified, and events that take place that have yet to be identified. One of those spectacular cosmic events is wrapped within the fine threads of the story that many identify as the birth of Christ.

People have been mentally programmed to imagine a baby in a manger when hearing the story of the birth of Christ. For today, that image needs an adjustment. Birth signifies something new has come, and babies are a mature stage of a collective group of cells – something new that has to do with a maturing process of the cells. The term Christ is used to identify a new condition or status in the blood (cells). This is why the terms Christ and blood are seen together in Scripture. Using this process of identification helps unravel the different character roles attached to Jesus. When the title Christ appears in text, it is to direct the attention to the cleansing activity that has become complete within the blood cells; there is no dead (contaminated) thing left circulating in His body. Once all stored up contamination is eliminated from the blood/cells, a person has "Christ."

The Star Dust must have existed in a group of people prior to the arrival of Jesus. It had become damaged or depleted to a point someone, in this case Jesus, needed to pay a visit to earth to get people back on track. The Star Dust can be depleted or eliminated by various actions, or lack thereof, of the person. Foods, sunlight, plants, bodies of water, hygiene and today we have toxins and chemicals, electronics, noise, all of which can deplete and eventually eliminate the Star Dust. The Star Dust is a must if the Soul of the person is going to exit this planet and make it to its next destination (Heaven). The intricate details of the journey of Jesus spell out a profound answer for us today.

Jesus takes on a character role of one who lives a life on this earth and takes on some challenging and sometimes unexplainable situations. The life story of Jesus is to show others that while painful and sometimes deadly physical situations can arise, the steps to obtain or increase the

Star Dust in the blood are what bring you out of dangerous situations.

What does Star Dust blood look like in terms of manifestations today? The Star Dust lies beneath any scientific discovery at this point. An outward display of a person's countenance and collective physical features will be noticeably different than what are common today. This is reflected in Scripture when reference is made to someone being a Hebrew. David is a good example. When Star Dust blood is present, the body will also be well, without the assistance of medications.

Here is where I John 5:12 should be considered: *The one who has the Son has life. The one who doesn't have the Son of God does not have life.* (HCS). The person who applies the collective techniques displayed by the life of Jesus, has the Son. The Star Dust results in qualification as a Son and that Star Dust life (Son) is a result of an attribute of God (Energy). Compare this to a child who looks very similar to its biological father. That child is not the father but displays a near likeness to him.

CHAPTER V
HOW OLD?

If Noah lived to be 950 years old, and many others lived into the multiple hundreds of years according to Scripture, why do people not have the ability to accomplish this today? I agree, 950 years of existence on earth sounds like a daunting task, particularly given the condition of the earth and the systems in place today. In addition, I doubt Social Security benefits would continue for this duration of time!

Isaiah 65:20: In her (a reference to a new heaven and new earth prophesied in verse 17), *a nursing infant will no longer live only a few days, or an old man not live out his days. Indeed, the one who dies at a hundred years old will be mourned as a young man, and the one who misses a hundred years will be considered cursed.* (CSB) (emphasis and description added)

If a person dies before reaching 100 years of age they are classified as being in their years of youth and have genetic imprints (cursed) that have likely contributed to their cause of death. Did ancestors leave behind curses for all who are on earth today? Yes, no question. If you have inherited disease, that is considered a curse. Sheds a little different light on the terms blessing and cursing to look at it this way.

It is rare for a person to reach 100 years, and survival to age 50 or 60 seems to be a challenge and certainly not doable without a prescription drug or two. Life-threatening health issues strike earlier in life with every generation. The verse in Isaiah tells me this is not how God intended life on earth to be. The curses must be removed before an age beyond 100 is attainable.

One might question why the mysteries of blessings and curses, the governing control the planets have over us, and how to live in accordance to their command were hidden from knowledge for so long. Part of the answer would be that it is how life was to play out.

Colossians 1:25-26: I have become its servant, according to God's commission that was given to me for you, to make the word of God fully known, the mystery hidden for ages and generations but now revealed to his saints. (CSB) (emphasis added)

Many players were involved in the removal of the ancient mysteries that has contributed to the financial and health related mess the world is in today. From who sold the valuable information called mysteries; whose hands the information was placed in; to, who contaminated the truth? Many hands were involved in the works to hide or destroy the truth. More important are the instructions that are coming to light now that when diligently practiced, will bring forth healthy, happy, long living human beings.

CHAPTER VI
UNVEILING THE ANSWER

While my method of presenting some of the answers up front may seem backwards, it will simplify the understanding of the chapters that follow to have a targeted conclusion in mind while you proceed. Admittedly, the recipe that produces the conclusion may not be simple to follow, particularly if you are not familiar with energy, science or interpretation of parables. Allow time for the information presented to soak in.

The human body is designed to be a conduit for chemical and electrical transport. We know this simply by the function of the heart. No chemical to fuel the electricity; no electricity present; no heartbeat. Electricity is produced in response to activity involving a fuel source. Food is not the most efficient fuel to power the electrical activity in the body. No one particular food added to or removed from the diet can be given an award for optimal function or dysfunction when it comes to the heart. Various diets are consumed everyday by people around the globe and yet the heart is still challenged with high cholesterol, stints, and failing heart valves. This concludes that something outside the realm of food needs to be had for proper function of the heart. While electrolytes in food are important, the most important fuel source for the function of the heart and health in general is helium.

Working my way through the maze of Scriptures I discovered some interesting facts that have led me to conclude that helium is a key player in the production of electricity necessary for the heart to pump and the streams of blood to flow properly. Helium, from what I understand, is in abundance under the earth's surface, in our atmosphere, and throughout the Universe, with large quantities making up the planets Saturn and Jupiter. Helium is not visible, nor does it have an odor. Helium gives off a yellowish glow, which would distinguish it from the natural gas it can accompany. Some interesting facts on helium:

* The reason why people develop squeaky voices after inhaling helium is that vocal cord vibration is faster when the gas that is around the cords has a lower density.
* Helium is the most abundant element in the universe.
* Helium emits a pale-yellow glow.

(TheFACTFile)

An interesting observation on this subject is the glow often seen around the head of Jesus. Jesus having a ghostly appearance and His ability to walk (float) on the water are the best examples of the presence of helium. So how do we as humans obtain this helium related ability to glow and to float? If Jesus did it, we should have the ability to do it as well, right? I am not convinced that people will begin floating through space once the helium is in proper balance and function. What I do see in the parable of Jesus walking on water is that there is a union between helium and water/plasma. The helium would move through the bloodstream while maintaining a connection to the plasma.

Helium is received by the body in a superior measure when proper meditation (called Sabbath in Scripture) is practiced. The helium may attract to our bodies at various times but for the greatest impact and an impact that will bring the healing necessary to wipe out the baggage ancestors left behind (the curses referenced earlier), there are specifics that must be applied.

Helium can have an impact on the body during a casual exposure, but many daily activities or environmental disturbances cause it to evaporate and not have any substantial influence within our own body. It comes and then leaves as fast as it came. People are too busy to notice. This process reminds me of how a helium balloon will slowly lose its ability to float. Too much helium in the body, though scientifically labeled as a non-toxic element, can cause damage to fibers or tissues in the body when in excess. Too much of a good thing can become a bad thing. A continuous taking in and releasing action must be in place. I call this taking in and releasing Love Exchange (Chapter XIII). The presence of helium can be felt as the body receives it, creating a sensation of being light-headed or overly relaxed. Makes one wonder if this is where the term "air head" originated.

Helium inside the body would accompany the plasma in the blood; vapor and gases seem to hang out together and vibrations record in or attract to water (plasma). When the helium and plasma team move through the veins, like a stream or river (two terms often used in Scripture), there is an electrical influence, causing a response that produces a component of healing and keeps the internal processes of the body optimally running with little or no interference. There are Scriptures that will be referenced later that mention the chest region of the body. This helium activity may be an answer to heart related diseases.

The circulation of the helium and plasma team, when in the proper portions and for the proper length of time, will remove harmful genetic imprints left behind by ancestors. The damaged DNA is repaired. This particular cleansing of the genetic imprints from ancestors only takes place <u>during the summer season</u>. At other times of the year, residue from infections encountered during our own course of life, the essence left behind from general noise, travel, foods, and so on will be cleansed from the blood through True Sabbath Meditation (Chapter XVI), a form of personal housekeeping. Many things that cross our path or that we put into our bodies progress and become a damaged link in the DNA chain if we are not diligent in our internal housekeeping. This does not mean that breathing in helium, like that used in balloons, will remove the potential harm surfing around in our body. The True Sabbath Meditation that results in a helium response is not and will not be available through a means of the medical (or any other) industry. This form of healing will not be copied by mankind. God will remain in first chair on this issue. An attempt to surpass the things God put in place and the risk of damage increases. The genetic imprints I refer to are scientifically identified as mutations that result in disease, for which the medical industry has no remedy.

It is important to note that not all DNA damage produces disease. DNA damage also spills out in altered body shape or size, skin tone, eye and hair color, texture of the hair, and so forth. While these various features may not be classified as damaging, the origin of human DNA that is identified as Hebrew, produces fair skin, petite individuals with blue eyes. (See, From AntiChrist to I AM by Harvest of Healing, LLC, published 2022.)

The body is designed to process the helium, not only causing the physical body to remain well and disease free, but also to release a form of the helium back into the environment. This is one of the hidden mysteries in the story of Adam being placed in the Garden of Eden. There is a slight semblance of Adam's physical body, being from the dust of the ground, exhibiting aspects of a plant that receives an element from the environment, filters it and returns it to the environment in a purified or more beneficial form. An act of photosynthesis. When outside treatments or therapies came on the scene (snake in the tree), the body no longer has an ability to continue its role in helping the environment remain healthy.

This photosynthesis is seen in the story of the woman with the issue of blood touching the hem of Jesus's garment as He passed through the crowd. When she touched His garment, He felt the power leave Him. He felt a noticeable shift. That end-product chemical that Jesus carried in His body cleared the harmful vibrations in the woman's blood. He supplied her what she needed for her blood to be healed.

Genesis 2:5-7, 15: ...no shrub of the field had yet grown on the land, and no plant of the field had yet sprouted, for the Lord God had not made it rain on the land, and there was no man to work the ground. But mist would come up from the earth and water all the ground. Then the Lord God formed the man out of the dust from the ground and breathed the breath of life into his nostrils, and the man became a living being. Verse 15: The Lord God took the man and placed him in the Garden of Eden to work it and watch over it. (CSB)

It takes human life to cause the sprouting of vegetation. The Breath of Life which comes from God contains vapor or mist, I think of clouds. That breath is not a breath like we would have here on earth but a vapor form that would come from the Heavens and is made up of various chemical gases. The Breath of Life that came from God would not be solely oxygen. Oxygen is not abundantly present in the Heavens and therefore is not what God would have delivered to Adam. Human life obviously needs a percentage of oxygen but what if the lifegiving form of what the human body needs contains a greater measure of helium, hydrogen or whatever other gases that are present in the Heavens? This concept may shed some light on why there were varying reports on the influence of the application of ventilators on COVID patients. Was oxygen being force into a COVID infected body causing a dangerous imbalance in chemical gases resulting in worsening symptoms or death? By servicing the respiratory issues with the current prescription for oxygen are individual health conditions worsening? The more you give thought to these types of things the more troubling they become!

It appears too much time is being spent on development of new potions (Rx) rather than researching the intricate details of what natural chemical elements are needed for life. Give the body what is needs and it will flourish. As an added nugget of insight, placing our feet in the soil is not an answer to this issue and can result in allergy symptoms, an irritation to the regulation of histamine.

With this picture in mind as you move forward through the chapters, keep in mind that there are ancient records that were removed, being absent from the printed Scripture we have access to today. More detailed information may come forth in the future when the ancient hidden away information is recovered.

CHAPTER VII

GOVERNING POWER

The cyclical rotation and positioning of planets will dictate the natural order of the physical body and all things on the earth. The evidence of its governing power may not be evident at any one moment but eventually the consequences from the power will come to fruition.

Saturn and Jupiter both contain significant amounts of helium, along with hydrogen. The gaseous vapor of helium encounters the physical body, gravitates to the plasma and results in a cleansing or canceling form of activity. There is no better explanation for the level of power that exists literally above our heads than in Romans 13.

Romans 13:1-6: Let everyone submit to the governing authorities, since there is no authority except from God, and the authorities that exist are instituted by God.
So then, the one who resists the authority is opposing God's command, and those who oppose it will bring judgment on themselves. For rulers are not a terror to good conduct, but to bad. Do you want to be unafraid of the one in authority? Do what is good, and you will have its approval. For it is God's servant for your good. But if you do wrong, be afraid, because it does not carry the sword for no reason. For it is God's servant, an avenger that brings wrath on the one who does wrong. Therefore, you must submit, not only because of wrath but also

because of your conscience. And for this reason you pay taxes, since the authorities are God's servants, continually attending to these tasks. (CSB)

These verses refer to the governing authority of the Heavens over human life and the influence it has on the physical aspects. The Heavens are suspended above every human being and the verses do not refer to any zone or location, so the influence of the Heavens applies to everyone on earth no matter their ethnicity. This line of thought initiates an investigation into the popular idea that the Abraham, Isaac and Jacob bloodline is the only ticket to the great hereafter. Dipping a little deeper into the "Chosen" (Matthew 22:14) bloodline, here's what I discovered: These three men are symbolic of 1) righteousness (Abraham); 2) he will rejoice (Isaac); and 3) supplanted; to follow closely (Jacob). Interpretation: A righteous life that produces an internal aspect of rejoicing when the laws of God/the Heavens are followed closely. This interpretation opens the door for any ethnic group, color, size or shape of person to have an ability to qualify as a member of Abraham's bloodline. Yes, this means that not just those who are of Jewish descent or who live in Israel can be participants. There are specific laws to be followed to qualify for the "Chosen" rank.

The "it" referenced in the verses is the power that the planets and their positions hold. The position of the stars and planets, the rising and setting of the sun and moon all have a governing influence over life. Follow the rules of the cosmic activity and you are safe from harmful influence. The verses are not referencing those in positions of power on earth, although the positions recognized here on earth reflect the governing order of the cosmic activities and influence thereof.

Amos 8:4-5: Hear this, you who trample on the needy and do away with the poor of the land, asking, "When will the New Moon be over so we may sell grain, and the Sabbath, so we may market wheat? We can reduce the measure while increasing the price and cheat with dishonest scales." (CSB)

There appears to be a form of eye in the sky that monitors our activity. New Moon and Sabbath days are not to be a time when money exchange that results in a profit is taking place. Nothing goes unnoticed and consequences would likely ensue. You are being monitored and influenced by cosmic activity whether you realize it or not. This insight may shed some light on the root issues of poverty.

Night sky activity communicates with the physical body in a beneficial manner different than the daytime hours. It is as though the planets release gases or other chemical compounds that result in some form of influence to the body. When the physical body is in a state of quietness or resting it has a better capability of receiving the signals given off by the planets.

When you are indoors there is less incident of overpowering signals that could afflict your body. Being indoors by sunset may be a good rule to follow. Protect the body from over-exposure of the cosmic activity by covering your body appropriately. No exposed arms, no holes or tears in clothing, no sheer fabrics, and so on. Appropriate attire is business or business casual, suits for men, dresses for women during the daytime hours. Evening or nighttime attire should include nightwear that appropriately covers the arms, core of the body and the thighs. Appropriate attire assists the body in receiving and protecting the activities in conjunction with the Heavens.

CHAPTER VIII
LIKE A DIAMOND IN THE SKY

Being made up of several planets and numerous stars that twinkle during the night, the Heavens hold more influence upon human life than what has been or even can be shared. Ancient Myths from various regions and people groups often refer to acts of nature, planets, thunderbolts and movements in the realm of the sky. If the voice of God is thunder, then why has the importance of Heavenly activity been taken off course?

Matthew 2:1-2: After Jesus was born in Bethlehem in Judea, during the time of King Herod, Magi (Wise Men) *from the East came to Jerusalem and asked, "Where is the one who has been born king of the Jews? We saw His star in the East and have come to worship him."* (NIV) (description and emphasis added)

Most readers are familiar with the story from Matthew 2. There is evidence in these verses that activity displayed throughout the Heavens not only has an influence but also produces a message. Note how the star is referred to as "His" star. I read studies a few years ago that suggest "His" star was the planet Jupiter. If so, this is an interesting fit to the other facts that are presented in upcoming chapters.

There is a category of people with the title of Magi or Wise Men that know how to determine what these Heavenly messages are saying. The Star of Bethlehem speaks of the influence from the Heavens that will lead to salvation, a rescue from health disruptions. Reading the messages displayed throughout the sky seems to be a skill that has become lost through the sands of time.

The activity in the Heavens has been taken for granted and in some cases completely ignored simply due to a lack of information or formal education necessary to understand or recognize its power. There are many references to the Heavens in Scripture. The understanding and mental perspective of the Heavens in some instances has been taken off course. Within the textual attempt to create the written stories we see in the Holy Bible, there can be references that are at times tricky to comprehend. With that, the Heavens have come to be understood as a place where angels hangout, and the soul of the dead is gathered after physical death. While some of this idea may have truth in it, there is also a broad door open for misconception of what the Heavens are and what they do.

There is a way to a more harmonious life, life will simply need to adjust before we can receive the necessary signals and elements (gifts) from above that results in healthy people. When mankind cannot receive and in turn give off the signals necessary to sustain life, all involved suffer.

CHAPTER IX
RUMBLINGS ABOVE

Matthew 24:29-31: Immediately after the tribulation of those days: The sun will be darkened, and the moon will not shed its light; the stars will fall from the sky, and the celestial powers will be shaken. Then the sign of the Son of Man will appear in the sky, and then all the peoples of the earth will mourn; and they will see the Son of Man coming on the clouds of heaven with power and great glory. He will send out His angels with a loud trumpet and they will gather His elect from the four winds, from one end of the sky to the other. (HCS)

These verses are telling of activity that takes place in the sky through a means of moon and star activity/ energy. Clouds and moisture will change or remove an electrical (or vibrational) interference. To cleanse or purify something the common practice is to place it under running water. Similar concept as taking a shower or rinsing off fresh fruits and vegetables, etc. This is the activity taking place in the sky above us. Disruptive recorded vibrations are being removed that will allow for a clear communication between people and the signals the planets produce. Interruptions experienced through loss of cell phone or navigation services is a sign of this Heavenly housekeeping.

A trumpet is symbolic of an announcement of something new; a difference that is making an approach. This cluster of supernatural events will open doors for specific individuals who will usher in future events that will spin mankind back into a life reflective of the ancient practices.

Moon Walk

The natural rhythm and heartbeat of the galaxy in its entirety has been interrupted. Massive amounts of metal shaped to form rockets or satellites all have caused some level of electro-magnetic, static electricity, gobbledygook that has created more damage than what has or ever will be admitted, discovered or could be described.

Why is it that humans conclude that metal robots can be launched into outer space and not cause some form of electrical or energetic interference? Every action taken will result in a reaction. Has this been forgotten? While we are at it, we should throw in the human footprints left on the surface of the moon. If mankind had been designed to live on or periodically visit the moon, mars, or wherever else NASA thinks they should or can visit, then the initial acts of creation (God, Big Bang, whatever you choose) would have put life on those planets long ago. There is a reason there is no physical life on those planets. Do people think God forgot something during creation? The planets we enjoy watching in the sky have assignments outside of human occupancy. It only makes sense that for the benefit of mankind as a whole and particularly the future generations, to keep our feet on our own turf. Floating around in the darkness of the galaxy is a disruption in the powers and energies it contains that help maintain life in all aspects on earth. The sign has been posted: "No Trespassing"! Research and discoveries should be conducted from earth without harmful invasion of outer space.

The moon signals were troubled when mankind made their invasion by setting foot on the moon June 20, 1969. Why is this date important? June 21 was and still is the Summer Solstice. While the depth of what takes place during the seasonal shifts is much more than what the human mind would have the ability to comprehend, the disruption that took place when mankind invaded the moon's space created a situation where the moon's signals that benefit the brain and cell communication in the body were disrupted. These moon signals were and are today, necessary for optimal electrical function of the body. Without the proper signals the ship will be taken off course! For many reading this book, I doubt there is a need to flip back in time to review details of the events, maybe I should call them shipwrecks, that took place in the 1970s that spilled out its share of messes and a few things gone wild!

It would be interesting to determine how many brain related diagnosis or illnesses were had after this 1969 event, and followed thereafter for the next 10 or 20 years. Would this lack of Heavenly signal be a root cause for migraines, bi-polar, schizophrenia, dementia, Alzheimer's or those mysterious undetermined cause of deaths? My extended family suffered the sudden loss of two family members during the 1970s, both of which had a fringe of mystery attached. A cousin was found dead in his college dorm bed, no evidence of any type of foul play, etc. His death was a fluke incident – so they say. All the "hurray" attached to this praised moon walk may very well have some "oh dear!" attached to it as well. The things humans do to grab a spot for their picture to be in the history book without having the full knowledge or even considering the negative impact their actions could have on innocent lives forever.

Luke 19:40: He answered, "I tell you, if they were to keep silent, the stones would cry out." (HCS). This verse is clear, rocks contain elemental forces that emit an essence into the atmosphere. We can only imagine the level of invisible force the moon or any other rock comprised planet contains.

Isaiah 14:3-17: When the Lord gives you rest from your pain, torment, and the hard labor you were forced to do, you will sing this song of contempt about the king of Babylon and say: how the oppressor has quieted down and how the raging has become quiet! The Lord has broken the staff of the wicked, the scepter of the rulers. It struck the people in anger with unceasing blows. It subdued the nations in rage with relentless persecution. All the earth is calm and at rest; people shout with a ringing cry. Even the cypresses and the cedars of Lebanon rejoice over you: "Since you have been laid low, no woodcutter has come against us." Sheol below is eager to greet your coming. He stirs up the spirits of the departed for you – all the rulers of the earth. He makes all the kings of the nations rise from their thrones. They all respond to you, saying: "You too have become as weak as we are; you have become like us! Your splendor has been brought down to Sheol, along with the music of your harps. Maggots are spread out under you, and worms cover you." Shining morning star how you have fallen from the heavens! You destroyer of nations, you have been cut down to the ground. You said to yourself: I will ascend to the heavens; I will set up my throne above the stars of God. I will sit on the mount of the gods' assembly, in the remotest parts of the North, I will ascend above the highest clouds; I will make myself like the Most High. But you will be brought down to Sheol into the deepest regions of the Pit. (HCS)

These verses give a good description of the trauma the Heavens have experienced and their recovery after the trauma is cleared away. Mankind disrupts many creations that should be left as a mystery, left for God and God alone to know the inner workings thereof.

CHAPTER X
THE CELESTIAL DANCE

Planets and stars dance through the sky providing a picturesque display that has profound meaning. Thousands of years ago this display was the only means for determining the time of day and the seasonal patterns for planting crops and so forth.

The Babylonians get credit for developing the 12 Astrological Signs and the Ancient Greeks take the credit for naming the 12 Star Signs after the constellations and linking them to dates based on their alignment with the sun's orbit. The Horoscope we know today gets no credit as it was not at play during the development of the Star Signs, nor used in Greece.

According to the Nationalgeographic.com website, the development of the 12 Star Signs was used as a means to converse with the gods. I interpret this as being a way to foresee acts of nature that would have impact on the daily life such as planting/harvesting of crops. The reference to gods can be a descriptive word for a powerful influence that comes from the Heavens. That influence becomes attached to a name (Zeus, Apollo, etc.) and put in the category of a god because the actions are a result of Heaven authored energies, some that today we would identify as a seasonal event. The complete definition of gods is an entirely separate ballgame and a more in depth look at the roles they play will need to be addressed at a later date, lest we end up on rabbit trails.

What became interesting is when I noticed the similarity in Zodiac signs/symbols and animals referenced throughout Scripture, along with the sacrifice any one animal might represent. To clear up some old beliefs, a sacrifice back then and for today is indicative of something you give up that would not necessarily be by choice; willing to do it but not pleased about it. Somewhere along the line of history the definition of sacrifice became contaminated and resulted in certain people groups literally taking the lives of others as a form of offering to the gods. This should have never taken place. Instead, one might sacrifice participation in a family or sporting event, or there may be occasions when you sacrifice not eating at a restaurant due to dietary restrictions. Basically, you give up something that may be valuable or important to you.

Details on what each sign may represent can be found easily on the internet. To be clear, I am not suggesting a daily horoscope reading or other mystic related practice. What I am suggesting is that the animals or persons (i.e., twin or maiden) referenced in Scripture may be indicative of specific roles a person born within that sign may have while here on earth. A Scripture reflective of a Zodiac sign may shed light on why specific individuals go through a more strenuous situation while others seem to have no problem with the same or similar situation at all. For example, an Aries/Ram or Taurus/Bull may have more challenges with blood related issues which could be reflective of the Scriptures that tell of the sin sacrifices involving goats or bulls. It appears Scripture does reference the Zodiac, it is just not forthright about.

Leviticus 16:8-10: After Aaron casts lots for the two goats, one lot for the Lord and the other for azazel, he is to present the goat chosen by lot for the Lord and sacrifice it as a sin offering. But the goat chosen by lot for azazel is to be presented alive before the Lord to make purification with it by sending it into the wilderness for azazel. (HCS). (Azazel: the strength to entirely remove)

There is a deeper mystery in the Zodiac signs being represented in Scripture than what there are currently words to explain. If you are familiar with Scripture, you will have no problem identifying each of the Zodiac signs therein. They are another way Heaven sends messages to earth.

- Aries is the sign of the ram (or goat);
- Taurus is the sign of the bull;
- Gemini is the sign of twins;
- Cancer is the sign of the crab;
- Leo is the sign of the lion;
- Virgo is the sign of the maiden;
- Libra is the sign of the scales of justice;
- Scorpio is the sign of the scorpion;
- Sagittarius is the sign of the archer;
- Capricorn is the sign of the sea goat;
- Aquarius is the sign for the water bearer; (a man with a water vessel on his shoulder).
- Pisces is the sign for the fish.

CHAPTER XI
COMMUNICATION FROM THE MOON

I admit to being no expert in the field of the various moon phases, but I have learned there are guidelines that should be followed to allow the maximum benefit and be in a continual position to receive the Love described in Chapter XIII, Love Exchange. Full Moon, New Moon, First and Third Quarter Moons all have their influence. There is far more to be learned about the various moon phases and the impact they have on life.

Waning Moon

The Waning Moon phase is the period between Full Moon and New Moon, when the moon's visible surface is decreasing.

Stepping outside of the governing laws of the Waning Moon has resulted in an extensive laundry list of health issues. Waning Moon activity influences the fluid in the body and when the laws are not followed, issues with fluids will arise. Compare this scenario with a house that has experienced anything from an overflowing toilet to flooding from torrential rains. This is just one example. Living Water (more on this in later chapters) must have a connection to these Waning Moon events.

The Waning Moon phase limits the consumption of specific foods. Personally, I have learned to avoid not only yeast but also cheese and honey during periods of Waning Moon. Consuming meat has very strict guidelines as well. Instead of running the risk of a potential interference with the goal I have of eliminating all DNA damage, I eliminated all meats and many other forms of protein altogether. This choice would certainly not be for everyone.

Using the Passover story as a reference point, the removal of yeast is to be practiced monthly beginning at Full Moon for a total of seven to eight days. This is the first week of the Waning Moon phase. Begin at 6:00 p.m. the evening prior to Full Moon and follow through until 6:00 a.m. the morning after the seventh day or the beginning of the Waning Gibbous phase. There is activity from the Heavens, described as the "death angel" in the story of Passover, that takes place during this weeklong period that influences the blood to a point that it causes yeast to have a negative impact and ultimately result in a form of death (dying of a disease; or born with a disability or birth defect; loss of the Star Dust/spiritual death). This weeklong fast from yeast is to be honored every month, not just one time per year.

If a yearly practice of removal of yeast was sufficient, it would be evident in a large group of people in the form of limited or no blood related disease. Every disease is connected to some form of contamination or alteration in the blood. To date, and to my knowledge, there is no large group of persons that are disease free. It is likely there is a long list of diagnosis that are related to the lack of proper Waning Moon yeast removal practice. Anything in the categories of blood, brain or nervous system disturbances would be at the top of the list. The scientific details of what yeast does to the blood and what takes place during Waning Moon is yet to be shared or discovered.

The Waning Moon phase is the time when the cells in the body are open, like the bloom of a flower. When the cells are open, they are more vulnerable to invaders and can more readily receive the cosmic message that initiates DNA repair. Mother Nature sends a reminder of this open cell stage by the display the blooms that are evident in spring and summer. An untimely influence or over-exposure to blooms can trigger a response from the cells that could result in an interruption to the natural process of cellular cleaning. The body, through the cells, reads symbols and exposure to open blooms outside of the designated season may cause the cycle of cells opening/closing to be out of sync with nature. Flowers should bloom in their season, not manipulated to bloom outside of their normal growing and blooming season. Anything outside of the normal blooming season could result in cellular miscommunication within the human body when exposed to blooms. This can be chalked up in the Mystical Wonders category. (Refer to the Chapter on Symbols in *Home-Made Answers for Cancer and Life Altering Disease* published 2024 by Harvest of Healing, LLC.)

The Waning Moon intensifies the sun and has an influence on the thigh muscles, region of the hips and reproductive organs, all areas of the body that are more often exposed to the elements due the choice of attire worn in response to the heat of summer. The more flesh you expose, the more physical symptoms you may experience in those regions along the road of life.

Waxing Moon

Waxing Moon phase is the period between New Moon and Full Moon, when the moon's visible surface is increasing.

Waxing Moon seems to be a little more forgiving allowing for the consumption of yeast and other foods that are restricted during various times of the monthly moon phases. It is advised to avoid the consumption of honey and yeast combinations, but it is permissible to consume them during the Waxing Moon phase, just not together. This would include the honey you may put in your morning beverage accompanied by the slice of toast you have for breakfast. So much for the Honey Whole Wheat breads. Until science can research the issue and come up with a plausible answer, it is best to avoid consuming the two together. (Leviticus 2:11)

Full Moon

Full Moon is when the entire surface of the moon is visible.

We have all heard of the Full Moon influence on wildlife and even hospitals will admit to there being an elevated need for medical attention during Full Moon. Oceanic activity is also influenced by the light of the moon. With the moon having an influence on the foods we eat, the time of day in which we eat, and the movement of the tides, there is certainly a whole lot of power going on out there. There seems to be waves of influence that occur and we will review that in more detail in the upcoming chapters.

New Moon

New Moon is when the entirety of the moon's surface is not visible.

New Moon is a marker for the beginning of cyclical activity that takes place in the Heavens and on earth. New Moon shares a level of importance with Sabbath, referenced in Amos 8:4-5 and II Kings 4:23. A more interesting note is the gender of a baby is determined by the phase of the moon at the time of conception. Such mysteries may tell us more about the numerous male descendants of Jacob.

Crescent and Quarter Moon

Crescent and Quarter Moons have an influence on the thyroid function. If you have a Hashimoto's diagnosis, the root cause may be an influence from the moon. Once more details come forth on how these partial moons are to be honored, it will open doors for a diminished occurrence of thyroid and possibly other health disruptions. Crescent Moons also play their part in the cleansing process of the body. Red is the color that sends signals to the blood. During a Crescent Moon it is advised to avoid wearing red. I will expand on this subject as more information becomes available.

Blood Moon

Acts 2:20: The sun will be turned to darkness and the moon to blood before the great and remarkable Day of the Lord comes. (HCS)

Recalling the fact that the earth will display the messages (or vibrations) recorded in the Heavens, what if this verse is telling us that the influence of the sun upon the people on earth will be at a level that it becomes harmful to the Soul, removing the Star Dust light that results in darkness? A good comparison might be how the color in fabrics become faded when exposed to extended periods in direct sun. And, what if a Blood Moon is the message of how the moon in general directs its attention to the blood?

Here is an interesting line of thought: I have never had much interest in vampire movies but weren't the actions of vampires triggered by the moon? Maybe there is a message behind Hollywood's vampire movies, of there being an uncontrollable influence directed to the blood that originates from the moon. Sometimes we just need to watch for hidden messages around us. Whatever the case and aside from all scientific reasoning behind Blood Moons, I sense the gods (for lack of a better description) are telling us our blood is powerfully influenced by the moon. Just to look up in the sky to see the results of an eclipse or the moon that has turned a pretty, reddish tone means nothing more than just a rare sight to see unless you look beyond the obvious. There is something more than just a grand display happening here.

If this all sounds strange to you, think of how the moon influences the tides and the activity of wildlife. A Farmer's Almanac is a good reference manual for the activity of planets and their influence on various aspects of life in general.

CHAPTER XII
STILLNESS

Psalm 46:10: Be still and know that I am God; I will be exalted among the nations, I will be exalted in the earth. (NIV)

Stillness is required for the body to not only receive but also process the communication it is receiving from the cosmic energies. If you are familiar with acupuncture, during and after the session the body requires a stillness to properly process the instruction it received. It would be of little benefit to receive an acupuncture therapy session and follow it with extensive activity. The body cannot process the signals received when it is overly active and will begin to drop the signals. A healthy level of stillness is required. Some translations of this verse refer to an act of discord or war needing to cease. This is not what the verse is directing the attention to. To "know" (God) is to experience an attribute of; reference to nations speaks of all ethnic people groups. To be still means just that, quiet yourself; remain in a state of being calm.

Obviously, this is not an invitation to become a couch potato. What I have witnessed and experienced is strenuous exercise can result in an unhealthy form of muscle stiffness or muscles feeling tense. It is possible this uncomfortable muscle issue is in response to a chemical reaction in the body. Somewhere within this equation could be an answer to cardiac issues. Some have read my statement before "I stay active. I do not 'exercise'." My body does not respond well to strenuous exercise, and I have never had a heart related health issue. My maternal grandmother lived to be 107, never did any form of formal exercise, and died as a result of her age. Like my grandmother, I stay active. I do not exercise.

Another form of becoming still is to stop talking, texting and so forth. No audio of any type is best for the body. It is amazing what the interior of the body can do when it is allowed to be in an atmosphere that is free of electrical disruption.

CHAPTER XIII
LOVE EXCHANGE

Vapor from clouds carry helium from the Heavens into our atmosphere where it is available to be absorbed by the physical body. The helium plasma combination circulates throughout the body. The body releases a resulting essence from this process back into the atmosphere. The earth must need a refined form of the helium for its vitality. Otherwise, the earth wouldn't necessarily need humans. God has a reason for everything. Adam, having been made from the dust of the earth, plays a key role in the purification or modification of the helium that was a benefit to the vegetation around him. This is comparable to the process used by trees to convert carbon dioxide to oxygen, the process of photosynthesis, so is the Love Exchange process between mankind and the earth.

If a person wants to be a part of God's Love, then they must learn how to be in the position to receive that Love and in turn, how to distribute it to the environment around them in a beneficial manner. Adam received the Love, a Cosmic Energy from the Heavens (through God's "breath"), processed it in his body then emitted the modified form of that Love to the vegetation and animals around him.

There is a chemical process involved in the receiving and releasing of Love that causes a physical sensation within the body. The best way to describe it is a sensation of champagne bubbles rising to the top of a glass. The helium, or a portion or form thereof, attaches to the plasma and in turn, this helium influenced plasma circulates throughout the body conducting a wash cycle that erases harmful vibrations present in the blood.

With the emphasis Scripture puts on the consumption of unleavened wheat bread, gluten protein must be a player in this chemical conversion cleansing process and/or the development of new healthy cells that coincide with that of a Hebrew. The role gluten protein plays is not yet clear but I will share what I have learned.

The New York Times printed an article in late 2023 or early 2024 about how the chemical structure of gluten changes when bread is left on a counter overnight. I do not have the name of the article or the date it was printed. This change in the chemical structure of the gluten would, in part, explain the reason Scripture instructs that bread (and food in general) should be made fresh daily. Many European countries serve fresh breads made daily. Sliced and day-old breads found at a typical grocers or restaurants would have less or possibly no benefit when it comes to the chemical compounds used to clean out damaged DNA and toxic blood.

Love takes on a whole different meaning in a spiritual aspect than what has been engrained into our minds and labeled as love. Having a rush of oxytocin, the love hormone used to describe a feeling of infatuation or lust, does not provide any form of healing process within the interior of the body and I doubt it provides any resulting benefit to surrounding vegetation. An emotionally satisfying feeling does not qualify as an internal healing. I Corinthians 13 provides a long list of what the Love from Heaven is and what it produces. People's actions today are far from what is described in I Corinthians 13. If the accepted definition of love was the superior definition, then the actions and reactions of people would not be what they are today.

Position Yourself

There are daily activities that can be done to draw the Love to your blood. The less exposure to environmental chemicals and electrical toxins, particularly static electricity, the better. Static electricity is not beneficial at any level to the health of the cells. Wear pastel colors and eliminate all synthetic materials. Synthetics are a source of static electricity. Quiet your activity level (more on this in Chapter XVI, True Sabbath Meditation). When it comes to eating, less is more. Limit the mixture of various food groups; consume no stimulants (caffeine, spicy foods, etc.). Some food groups, though they may taste good and pass through the body seemingly without impact, will cause excess energy to gravitate to the digestive system to process the meal eaten and contribute to disruption of in the chemical balance necessary to complete the Love cycle. Food intake should be in small portions throughout the day. This is what Ezekiel describes.

Ezekiel 4:10: The food you eat each day will be eight ounces by weight; you will eat it from time to time. (HCS)

CHAPTER XIV

TIME WILL TELL

Nighttime Lullaby

Nighttime can bring some interesting events, from children being frightened of the Boogeyman, to being startled by unidentified noises, all seem to just go with the territory of being in darkness. Although, according to Job, we should not be too quick to dismiss everything we may experience in the night:

Job 33:14: For God speaks time and again, but a person may not notice it. In a dream, a vision in the night, when deep sleep falls on people as they slumber on their beds, He uncovers their ears at that time and terrifies them with warnings, in order to turn a person from his actions and suppress his pride. God spares his soul from the Pit, his life from crossing the river of death. (HCS)

Proverbs 6:9-11: How long will you stay in bed, you slacker? When will you get up from your sleep? A little sleep, a little slumber, a little folding of the arms to rest, and your poverty will come like a robber, your need, like a bandit. (HCS)

When the blood lacks Star Dust, the components necessary to conduct the cleansing of the blood is absent. If you can sleep six to eight hours without interruption, your natural Star Dust housekeeping may be shut down. When the Star Dust is present and is at work, it is not uncommon to experience sleep interruption. The verses in Proverbs are speaking of a state of being totally unaware (of what your internal circumstances are). Poverty is a descriptive term for those in ill health or lacking a proper Star Dust level, not the value of your bank account or assets.

The medical industry has placed great emphasis on the necessity of sleeping an average of seven hours each night, depending on your age. This is impossible for those with active Star Dust, and to attempt to gain the goal set forth by the medical community (and likely Big Pharma) could end in only frustration and a reduction in required Star Dust activity to keep the DNA protected from damage. Sleep disruption is an indication the body is cleaning out contamination it encountered during the events of the day. A storm of internal activity is at play.

Mark 13:34-36: It is like a man on a journey, who left his house, gave authority to his servants, gave each one his work and commanded the doorkeeper to be alert. Therefore be alert, since you don't know when the master of the house is coming—whether in the evening or at midnight or at the crowing of the rooster or early in the morning. Otherwise, when he comes suddenly he might find you sleeping. (CSB)

Mark describes times when you have departed healthy genetics (house; where the family should be) and are unaware of what may go on while you are away from your intended place of living. Some form of alarm system needs to be in place. Your body needs to have some way to tell you something is brewing that can result in trouble. With the mention of evening, midnight, crowing of the rooster and early morning, it is apparent these are the times when the body will exhibit a symptom in response to a potential health danger. Any symptom experienced will reflect your personal genetic issue, organs involved or not. The "master of the house" would be the current issue being addressed within the body such as an infection you encountered, a mastering genetic imprint or an essence left behind from a food or other form of exposure that is stirring up trouble.

While awake the heart rate and respiratory rate increase. This rate increase must have a relevance in the process of eliminating the potential danger. Forcing sleep during these times could create a situation where the body is unable to capture the opposing vibration it has identified and eliminate it. The longer the harmful vibrations linger in the body, the greater the risk becomes for those vibrations to cause damage to DNA.

I Thessalonians 5:7-9: For those who sleep, sleep at night and those who get drunk are drunk at night. But since we belong to the day, we must be serious and put the armor of faith and love on our chest and put on a helmet of the hope of salvation. For God did not appoint us to wrath, but to obtain salvation through our Lord Jesus Christ. (HCS)

We are commissioned to act and trust that the Love (from the Love Exchange) is protecting us, specifically in the region of the chest. This coincides with Scriptures that refer to the heart. (i.e., Create in me a clean heart; renew a proper Spirit (Love energy influence) within me.) Sleep and drunk are terms used to indicate being unaware of your circumstances while being under a harmful influence, such as an infection or other threat to the cells/DNA. Helmets are protection for the brain and the crown of the head. This would be a reference to mental health and can have a connection to Scriptures that speak of the madman. Lastly, God did not send us to earth to be beat up by suffering and disease. There is a plan of rescue put in place evident to us through the life events played through the character role of Jesus.

The Impact of Midday

The hours from noon until 3:00 p.m. hold mysteries pertaining to some form of influence or the lack of a protection to the physical body. A covering of protection seems to disappear. We see a hint of this in the story of the crucifixion when:

Matthew 27:45: "From noon until 3:00 in the afternoon darkness came over the whole land. About 3:00 in the afternoon Jesus cried out with a loud voice – "My God, my God, why have you abandoned me?" (CSB) (emphasis added)

In this verse, an important element (aka God) abandons the atmosphere coupled with an action that results in darkness; the physical body encounters an impact. Darkness in this reference would be the removal of Star Dust light from the blood plasma. Why do I select the plasma? In the story of the crucifixion when the soldier pierced the side of Jesus water poured out, indicative of an issue with the blood plasma. A separation from what sustains the body occurs in this story. Is the intensity or condition of the sun at play during these hours in a fashion that causes the Star Dust light to fade?

Luke 23:44-46: It was now about noon and the darkness came over the whole land until three, because the sun's light failed. The curtain of the sanctuary was split down the middle. And Jesus called out with a loud voice, "Father, into your hands I entrust my spirit." Saying this, he breathed his last. (CSB)

There is not only an intense change around the hour of noon and lasting until 3:00 p.m. but that change influences the inner most being of the human body, the sanctuary. "Down the middle" would be the region on the chest, or core of the body. Again, as mentioned in the prior verses referenced in I Thessalonians, this directs our attention to the heart. This activity concludes with "into your hands." Hands reference work; what you are known for, a skill. The functioning aspects that take place within the region of the chest are controlled by the cosmic energies. The vibrancy of the Soul must be entrusted to the Cosmic Energy (the works of God).

Psalm 91:5-6 tells us that there are "arrows that fly by day", "plagues that stalk in darkness" and "pestilence that ravages <u>at noon</u>." A pestilence is a deadly or overwhelming disease.

Two things could be at play in these situations: 1) either people have lost their ability to safely weather the impact present during the hours of noon until 3:00 p.m., or, 2) the patterns of nature are telling us we need to limit exposure to the outdoors and excessive activity during these hours because of an influence that occurs on the interior of the body in the region of the chest. Plagues that stalk in darkness are damaged DNA strands that are a result of lack of Star Dust light from the blood. An interesting study would be of how many heart attacks or cardiac distress incidents occur during these hours.

Lunch

Verses from the Old Testament present that "bread" is to be eaten in the morning and in the evening. There is no mention of lunch. This caused me to do a little research on where the institution of lunch originated.

The history of where and why lunch began appears to vary dependent upon the country. For the United States there were a few articles that presented interesting information on the subject. Lunch at one time was considered the workman's meal. Prior to working away from home, lunch was not a common meal. President Roosevelt was in office from the mid-1930s through mid-1940s signing the Pure Food and Drug Act that later birthed the Food and Drug Administration. With the Great Depression and World War II at hand, somewhere in the mix lunch became a popular mealtime for all.

With the emphasis on the hours of noon until 3:00 p.m., lunch appears be a detriment to the interior process the body is attempting to conduct. The specifics, like most instructions that unfold from ancient text, of what happens and why food would be a problem is still being untangled. Eating a form of wheat bread in the morning (pancake or biscuit for breakfast), followed by fruit 1 ½ - 2 hours after consuming bread, yet before 11:00 a.m. is permissible. No food should be eaten between noon and 3:00 p.m. Waiting until after 3:00 p.m. to consume any meal was a part of the protocol I followed to escape a fast-approaching cancer diagnosis.

Twists and turns exist that must be ironed out with respect to lunch. I do believe lunch is a contributor to the interruptions that take place resulting in lack of or delay in the internal cleansing of the blood. There may also be seasonal considerations that need to be investigated.

Permission to Snooze

The busy American lifestyle has pushed aside some ancient practices that exist today in other countries. While visiting Spain and Italy it was apparent the culture honored a time of retreat from work granting a time of rest in the middle of the day. Many businesses, unless they were a restaurant that served lunch, would close from noon until 3:00 or 4:00 p.m., reopening for the late afternoon to early evening hours.

II Samuel 4:5: Rechab and Baanah, sons of Rimmon the Beerothitem set out and arrived at Ish-bo-sheth's house during the heat of the day while the king was taking his midday nap. (CSB) (emphasis added)

This Scripture may hold an answer as to why the siesta is popular in Mexico. This quick catnap that you may see many laborers participate in is not a sign of being lazy but a sign of following an ancient teaching that aids in the protection and refreshment of the Soul. Remove yourself from work and sun exposure and rest.

A Hat For Every Occasion

The verses in Deuteronomy shared below, cast a light upon noon and a connection to blindness, madness and mental confusion. The crown chakra is on the top of the head and should be in a position of being upright, although the chakra can become tilted and misaligned. Once again, referencing the story of the crucifixion, Jesus displays a crown of thorns around His head. This crown is a sign of discomfort to the region of the head or brow + the blood flow. If a person must be outdoors, especially during the hours from noon to 3:00 p.m., I strongly encourage the head being covered with a brimmed hat.

It is common for people who live in the middle east to cover their head. Again, this could be the sign of an ancient wisdom being practiced. If you have ever watched movies from the 1800's, whether of Royals or of cowboys, wearing a hat was quite common. Ever notice the umbrellas carried by many from Asian countries as they walk in the sunlight? From what I have witnessed many, particularly those who display an age of wisdom, cover their head with either a large brimmed hat or an umbrella. As a means for protecting our brains and as an aid to prevent issues with vision and mental health, I suggest implementing a hat into your wardrobe -- rain, shine or otherwise.

Deuteronomy 28:28-29: The Lord will afflict you with madness, blindness and mental confusion, so that at noon you will grope as a blind person gropes in the dark. You will not be successful in anything you do. You will only be oppressed and robbed continually and no one will help you. (CSB) ("Lord" is something greater or more powerful, higher ranking. Robbed speaks of the decline in the Star Dust light, which is our means of internal protection.)

Untimely Death

In the story of the Shunammite's son set forth in II Kings 4:18-37, the Shunammite woman's son was ill and sat on her lap until noon and then died. Like any mother in that situation, she got up and was determined to locate help. Her actions were questioned when she announced she was going to "the man of God" to get help for her son. Verse 23 states: ... "*Why go to him today? It's not a New Moon or a Sabbath.*" This statement makes it clear that days of Sabbath and New Moon were known for powerful healing; a healing that would cancel the order of death (damaged DNA) and resurrect the Soul through the replenishing of Star Dust. Note the reference to noon and death (of the Soul).

The sun must have such an intense signal that it can and will in the correct circumstances, eventually remove the River of Light (Chapter XVIII) – the light that is carried by the plasma.

Early Bird

I Samuel 29:10: So get up early in the morning you and your masters' servants who came with you. When you've all gotten up early, go as soon as it's light. (CSB)

Luke 21:38: Then all the people would come early in the morning to hear him in the temple. (CSB)

Gearing the mind toward health, in the verses above a master is what rules over your family tree, or the combination of components that construct the physical body. This would be the chain of DNA that is designed in a healthy sequence or genetic imprints, being damage to the DNA strands. There is activity in the cells that takes place at daybreak. The reference to the temple in Luke is referring to the body. To "hear him" reflects an activity in the atmosphere that can produce a message. Whether that message is audible, visual, or simply a thought that passes through the mind would be dependent upon the person and the circumstances. Regardless, there is important activity connected with the dawning of the day. Like the early bird who captures the worm, it would be wise to be alert and ready to receive.

John 20:1: On the first day of the week Mary Magdaline came to the tomb early, while it was still dark. She saw that the stone had been removed from the tomb. (HCS)

Daybreak brings about the evidence that what was holding a person in a death bound health situation has now been removed. Stones represent difficult things to remove or relocate. Sunday morning, noted as the first day of the week by John, following the True Sabbath Meditation, a new beginning has occurred in the body. More light is shed on this subject in Chapter XVI, True Sabbath Meditation.

Psalm 30:5: For His anger lasts only a moment, but His favor, a lifetime. Weeping may spend the night, but there is joy in the morning. (HCS)

By now we get the idea that some form of internal activity is going on that is initiated by a Cosmic Energy and can result in either harm or health.

CHAPTER XV
THE POWER OF CHANGE

If mankind is going to see change, mankind must be willing to change. Things are not going to change on their own. If change is not implemented, the average age of death will continue to decline.

There are events in Scripture that piece together a very interesting picture. I will explain the setting first as an aid for grasping the story line and then provide the Scriptures.

The previous chapters have revealed that blood plasma can receive a supernatural charge from signals and/or elements that originate from the planets (i.e., Saturn). Clouds are the transportation system for the helium. The transported helium makes contact with the blood plasma. In turn, this brings about certain events that take place within the body that result in the removal of toxic vibrations or debris that the physical body has encountered throughout a day or week, or even inherited from three or four generations of ancestors. This set of events, which I hope I have shortened to an understandable level, is all laid out in Scripture.

Scripture sets the stage quite well when it comes to taking care of the physical body. Two things must be brought to your attention: 1) People sought care through a "man of God" or through Jesus. Reference to a man of God is not pointing in the direction of a Priest or other church leader as we know them today. When the care of a physician was sought, the money became depleted and/or the affliction was still present or became worse (Mark 5:25-26). Physicians are also noted in Scripture as being those who embalmed the dead (Genesis 50:2). Now that point could spark some interesting conversations! 2) Jesus did not attend weekly religious gatherings. Does this point to the fact that Jesus and/or healing will not be found inside a church? That is for you to decide. In Luke 4, Jesus is noted as being present at the temple on a Sabbath where He read Luke 4:18-19. Again, our "body is the temple." These links in the chain of events can be indicative of activity in the body that takes place in response to the moon or a Sabbath, not necessarily a physical appearance at an organized event.

How do we get this physical and spiritual healing without participating in regular medical consultations and appointments or without participating in a religious group?

Jacob's Well

John 4:1-26 is the story of Jesus visiting with the Samaritan woman at Jacob's Well. Several details unfold in this story and need to be highlighted: 1) Jesus is worn out from His journey and sits by the well about noon. 2) Jacob's Well is a symbol of something deep within the body that has a connection to water; plasma. 3) The disciples travel to get food. Food and water are both refused by Jesus. Why? The Living Water (Star Dust enhanced plasma) has work to do outside of digesting food and supplies sufficient sustenance. Note that Jesus was sitting, resting after his morning activity. This Living Water has a spiritual connection to the Jacob blood(line). 4) The Samaritan woman has no husband, but previously had

5. The title master and husband are interchangeable, both referencing a source that has precedence over a household or family; a daily order of events. 5) Samaria references a people group of a non-Hebrew race. A non-Hebrew in this story would need to consume water because there is no practice in place that would provide the Living Water within. Those classified as being from a Hebrew race have the practice(s) in place and resulting internal resources that provide the hydration necessary for the body outside of the need for drinking water.

The reference of Samaritan and Hebrew are descriptions of a status inside the body. If your body requires the consumption of water, you are a Samaritan, if you consume no water and have the Living Water, you have graduated to a Hebrew status. These two terms, particularly Hebrew, are used throughout Scripture. From this portion of text we have: a source of naturally reoccurring hydration that is within the interior of the body; resting with no food or drink (water) at noon.

Officially A Royal

John 4:46-54 tell a story of a certain royal official whose son was ill. This royal official approaches Jesus and pleaded for the healing of his son as his son was about to die. Jesus was a bit perturbed by the royal official's lack of belief and informs him that his son will live. The royal official departs and travels back to his home where he learns that his son's fever had left. When informed that the fever departed the boy at 1:00 p.m. (the seventh hour), the royal official realizes this was the time of day when Jesus had said "your son will live." (Some translations state the hour as seven in the morning. This translation is incorrect.) There is a connection between the Living Water in the body of a Hebrew race and the hours of noon and 1:00 p.m. that produces a healing or relief from a form of infection. Living Water must be at play for death to depart.

It is beginning to appear as though the traditional lunch hour used for consumption of food has contributed to the rise in health issues, not to mention the added pounds that seem to develop as a result.

Gene Pool of Bethesda

John 5:1-16 tell a story of the disabled man who laid near the pool of Bethesda. The angel (a power from the Heavens) would move or stir the waters and the man lacked the ability to reach the water due to his physical condition. Thirty-eight years of physical impairment was experienced by this man. The number 38 may be indicative of the years endured having no weekly cleansing of blood/DNA in place – lack of participation in a regular Sabbath Meditation that resulted in his physical disability. After Jesus directs a few questions to the man, He simply tells him to gather his things and go on his way; problem solved. This encounter took place on a Sabbath. This story gives us the following pieces: 1) a pool is reference to the gene pool. This particular gene pool has a history of debilitating physical impairment dictating an inability to reach the condition the plasma needs within the body to produce a healing. This healing occurs in response to the Sabbath, a form of extended meditation described in Chapter XVI.

Thus far we have movement in the plasma between the hours of noon and 1:00 p.m. connects to a Sabbath and results in healing of the gene pool.

Turbulent Waters

In Matthew 14:22-33 Jesus and his disciples had been amongst a crowd and the meaning of their gathering had come to a close. Jesus instructs the disciples to go ahead of him by getting into a boat and make their way to the other side. Jesus dismissed the crowd and then retreats for some evening alone time. The disciples experience some troubling water related events as they were making their way to the other side being about a mile from land; wind and battering waves come on the scene and cause them distress. Around 3:00 a.m. the healing rescue in the form of Jesus appears on the water. This form is not immediately identifiable by the disciples. This ghostly form with the ability to float or, as Scripture states, walk on the water, approaches their situation of need and ultimately calms the disruption experienced by the disciples.

This parable is describing the challenges that will come against a person's physical body during a time when the health of the blood is threatened. Notice Jesus had dismissed the crowd, sending them on their way, and retreated to be alone (in the evening). To draw the helium to the plasma, being alone, away from excess noise and activity is key. The various vibrations created by talking, music, general noise, vehicle movement, etc. is disruptive. The blood plasma becomes disturbed (wind and waves) until helium, a floating vapor form, arrives and moves into its position with the plasma (Jesus floating on top of the water). Helium has a glow to it and can produce an act of floatation. Three in the morning brings the calming of the disturbing events.

The boat being a mile away from shore is reflective of how separating ourselves from the ancient practices, such as

Sabbath Meditation, will result in damage to DNA. We end up in an ocean of problems. Considering the distance of a mile, 5,280 feet, and the length of a DNA strand of 6 feet within each cell, 5280 divided by 6 equals 880. This gives us an approximate number of cells that need repair when turbulent health issues begin, and symptoms make their debut. Approximately 880 cells that contain damaged strands of DNA triggers the plasma to become disrupted, turbulent. Before having the ability to calm a health situation the DNA strands must be healed through the helium and plasma combination. When the DNA is not healed the mRNA process copies the damaged cell information, and the health issue intensifies. Every cell holds a strand of DNA and there are approximately 75 trillion cells in the body. (KnowyourDNA.com). Add years of cell multiplication and continued toxic accumulation to this turbulent water situation and the number of damaged DNA strands increases.

The blood plasma must contact an element of helium for the storm of health issues to calm down. Remember the reference to helium and vibration in the facts listed? Vibrations make up the genetic imprints and it takes an opposite or superior vibration to remove or cancel a vibration. Water records vibrations and if the helium provides a healing form of vibration, the movement of the helium plasma combo through the circulatory system will bring a total body healing. One dose will not do the trick. It takes several months to repair hundreds, if not thousands of feet of damaged DNA strands. A regular practice of Sabbath Meditation must be in place to maintain a healthy plasma and prevent genetic imprints from trickling through the genetic bloodline.

Definitions of the various named characters and locations are provided:

<u>Bethesda</u>: house + authority. An authority (cosmic) that rules over a house (group) of people. One house connects to another house – governed and controlled by a "father", executed by the Sons add up to be the mother/tribe (describes generations). To be good or kind; loving kindness (House of Mercy).

<u>Pool</u>: water confined to a specific location or perimeter; naturally formed, stationary. Gene pool; collection of recorded information from ancestors.

<u>Sychar</u>: hired wages (to be paid/receive a benefit); influence as though drunkenness (an uncontrollable influence).

<u>Samaria</u>: guard, give heed, observe; to exercise great care. A non-Hebrew race.

Dirty Laundry

The human body holds on to various things it encounters throughout each day. The beats to music, the essence from foods eaten, the sunlight we are exposed to, the grass we may walk in, everything. Some of these things are neutral in their influence, some beneficial, but many alter the Hebrew DNA. By the end of a week, we have a basket full of contamination or disruption in our body that takes up residence in the blood, like having the accumulation of dirty laundry. This accumulation of what I call sound bite chatter causes disturbances that lead to various diseases. After this accumulation of dirty laundry sits in the bloodstream for a period of time, it causes contamination, eventually making its way to the DNA strands and results in altered or damaged DNA. Today, people not only have the internal messes made through their own life, but also have an accumulation of messes from the past three to four generations. (Exodus 20:5-6). This pattern of continual accumulation of harmful vibrations, the dirty laundry contrary to the original genetic makeup (called Hebrew), is what will eventually spill out into altered facial features, uncontrollable weight issues, variations in eye, skin or hair color, and so forth. It would be like adding a teaspoon of sand to a glass of water each day. If you fail to dump out the sand eventually the sand will take over the water and you have no means of cleansing the internal body. The cells become full of toxic waste in the form of vibrations. You can take numerous supplements or a detoxing herb and they will not reach the arena where the toxic vibrations are doing their dance. The only way to reset the internal cellular frequency is to practice weekly Sabbath Meditation described in Chapter XVI, and to eat unleavened wheat bread.

With the load of inherited disease people have today, this would be a good place to insert: Thou shall not commit murder. There are multiple ways to murder someone and the meaning of this commandment may not have been as broad as it should have been. Several who have lost their life to inherited disease could add the phrase "My Ancestors Murdered Me" to their headstone.

The body responds to sound bites in cells but the display of what it hears is in a fashion that modern day medical treatment or diagnosis cannot administer an effective treatment or eliminate the cause. This is what took place in the man by the pool (gene pool) of Bethesda. The genetics from ancestors were so buried by his own daily exposure to various disruptions, the genetic laundry could not be cleaned. The good thing about this story is there is an element of Sabbath and Jesus involved. To be clear, Jesus is representative of a collection of actions that must take place to produce a healing, a form of rescue from what ails you.

During an internal bath of the body, the body will address the most recent contamination first. The more contamination the body has accumulated throughout the week, the more internal laundry there is to do. The body can only handle so much. When the accumulation becomes too much for the body to process in one week, the excess hangs out in the blood until the following week, and so on. If the internal cleansing process never catches up to the level of accumulation, after some time there is too much internal dirty laundry to handle, DNA strands become damaged, disease kicks in and your dirty laundry passes on to the next generation. This accumulation over time will result in a form of cancer.

Life-threatening health situations can become frightening simply because there is a factor of the unknown at play. It is never easy to predict how the body may react since every human body has a variation of multiple vibrations (genetics) that are causing health disruption. Knowing that whatever is taking place is a natural process is important. Whether you can weather the internal storm without medical intervention is dependent upon the issue. We must learn to properly interpret what the body is telling us. Anytime the body goes through a significant shift, the body will display some form of symptom simply because of the activity involved that is necessary to birth the change. Like labor pains during childbirth, it can be painful but at the conclusion you receive a new life.

CHAPTER XVI
TRUE SABBATH MEDITATION

<u>Sab</u>: Person engaged in direct action to prevent a targeted activity.

<u>Bath</u>: To immerse or wash one's body.

The Jewish practice of counting from the New Moon forward to determine the dates of Sabbath appears to have flaws. If this concept was the correct way to determine the Sabbath dates, many people would present as healthy, without a need or very little need for medical intervention. Healthy is not defined by a means of medical intervention, particularly the methods used that will eventually fail and the race to the cemetery begins. I suspect that in days past the counting from New Moon to determine Sabbath dates was a correct practice. What appears to be an issue is the introduction of the Gregorian Calendaring System by a papal decree in 1582. When a person in a position of authority steps in and formulates a law that interferes with a current practice, the response from the Heavens shifts. A new law will often overrule an old practice. A new law will not alter the natural order so no, the rising and setting of the sun or moon cannot be changed for more than brief periods of time anyway, according to Scripture. For actions that are taken by humans yes, the cosmic law will shift in response to an order issued by those in authority. Written documents have great power when executed by an authority figure. Thoughts arise as to whether the summer and winter solstice dates are correct, along with many other dates that may have encountered a shift with the adjustments made to the calendaring system.

For today, Sabbath is recognized by the Heavens as being on Saturday, the 7th day of the week and being named after Saturn's Day, symbolic of helium which is an element that makes up a portion of the planet Saturn. In the current day, it would be quite challenging for people to follow a counting from New Moon system to honor a Sabbath. With common work schedules being Monday through Friday, and sometimes Saturday, it makes honoring a New Moon calculation style of a Sabbath impossible for some. For Sabbath Meditation I follow the example of the death and resurrection of Jesus.

Friday evening through early Monday morning is spent in varying degrees of isolation. You are basically in a tomb or hidden state. Eat your Friday evening meal at home with family members only. Scriptures describe Jesus as being alone or having dismissed the crowds. Less physical activity leaves more room for God Energy activity. Obviously if you have small children or elderly parents that you supervise, being alone may be impossible. Do the best you can.

Saturday morning is met with a shower and getting dressed in clothing of a pastel color, or combination of colors. No black, gray, navy, white, or other dark colors; no synthetic material. A small serving of concord grape juice is permissible on Waxing Moon Sabbaths; no food if your health condition allows for a brief fast. A light breakfast of unleavened bread and concord grape juice is to be had on Waning Moon Sabbaths. Once the initial juice or bread/juice consumption is had, consume no additional food until after 1:00 p.m. The less you put in the body, the more influential the cleansing process will be.

No household chores, no cooking, nothing that involves heat, no television, telephone or internet, no socializing or visitors, no shopping or traveling, no sexual activity. Any form of activity could interrupt the ability to receive the required level of helium to bring forth a sufficient healing, or to erase the genetic imprints that are triggering a disease. A physics action takes place, joining of Spirit and Flesh (union).

Select a comfortable place to sit or recline. Within approximately 30 minutes after entering the comfortable, quiet space, you will feel relaxed and eventually sleepy; experiencing a light-headed feeling is not uncommon. Remain in this relaxed state until after 1:00 p.m. It is quite possible you will be short of energy the remainder of the day.

Sunday morning you may introduce breakfast and implement moderate daily activity. The body is going through a transition and any excessive exercise or other harsh activity can result in a setback of your efforts. Eat as you normally would on Sunday and on Monday morning normal activity may be resumed.

Isaiah 58:13-14: If you keep from desecrating the Sabbath, from doing whatever you want on my holy day; if you call the Sabbath a delight, and the holy day of the Lord honorable; if you honor it, not going your own ways, seeking your own pleasure, or talking business; then you will delight in the Lord, and I will make you ride over the heights of the land, and let you enjoy the heritage of your father Jacob. (CSB)

The Sabbath Meditation process is similar to the more recognized meditation atmosphere but does far more than the typical meditation simply because you are in a position to receive in accordance with the laws that govern the Heavens. Sabbath Meditation takes more time and focus than the typical 20 to 30 minute time of solitude.

Why is 1:00 p.m. important? Refer to the story in John 4 of the boy whose fever left him at one in the afternoon. After 1:00 p.m. I suggest eating something that is within the cleansing protocol of: unleavened bread, concord grape juice, fresh fruits or vegetables. No heavy meals.

Within the stories of the Samaritan woman at Jacob's Well about noon, Jacob's Well is a reference to water deep within (the body); the man by the pool of Bethesda (gene pool); and the boy's fever leaving at 1:00 p.m. there are highlights of the process required for a Sabbath Meditation that will eliminate genetic imprints.

Will the harmful imprints all be removed in one Sabbath Meditation? Doubtful. It may take one person 10 Sabbath Meditations sessions to eliminate their genetic imprints and another person it may only take 5. It all depends on what has been recorded in your genetics and how disciplined you are at following what is required to clean the harmful imprints out.

I Corinthians 15:41-42: There is splendor of the sun, another of the moon, and another of the stars; in fact, one star differs from another star in splendor. So it is with the resurrection of the dead: Sown in corruption, raised in incorruption; ... (CSB)

As previously mentioned, the inherited genetic imprints are removed during summer season only. Nevertheless, as a rule of good internal housekeeping, it is wise to continue a weekly Sabbath Meditation practice. Often what we encounter on a daily basis causes a disturbance within the body that does not become evident until years later. For example, static electricity does more harm to cells than what is believed, along with the long list of toxins, EMF/EMR, and so on. Suggestion: To help reduce the negative influence of static electricity, fill a fine-mist spray bottle with water and a drop or two of an essential oil of your choice and spritz the air throughout your home.

Unless you are experiencing a measurable fever at the time of a Sabbath Meditation, evidence of any change or shift inside the body may not be evident for several weeks. The body must work at removing the contamination captured and any exterior evidence of the toxic removal may be weeks ahead. Therefore, the less exposure to harmful environments during the week, the less self-imposed contamination, the more likely the genetic imprints can be addressed during summer Sabbaths.

A child should reach the age of 12 before beginning a regular Sabbath Meditation practice.

CHAPTER XVII

HOLY SPIRIT

<u>Holy</u>: Perfect in goodness and righteousness; an aspect of God.

<u>Spirit</u>: The non-physical part (of a person) which is the seat of character; the formation of the definitive or typical elements in the character (of a person).

<u>Pentecost</u>: Fiftieth

Adding in another important point in the influential response to helium is what Scripture calls being filled with the Holy Spirit. When the body encounters an element of helium that is a result of a True Sabbath Meditation, the body can feel relaxed, maybe a form of lethargic or sleepy; light-headed. The gases inside the body are mingling and shifting. I believe this influence of the helium is what Scripture at times calls drunkenness, an influence that is uncontrollable.

Acts 2:1-4; 12-13: When the day of Pentecost had arrived, they were all together in one place. Suddenly, a sound like that of a violent rushing wind came from heaven and it filled the whole house where they were staying. And tongues, like flames of fire that were divided, appeared to them and rested on each one of them. Then they were all filled with the Holy Spirit and began to speak in different languages, as the Spirit gave them ability for speech. Verse 12-13: They were all astounded and perplexed, saying to one another, "What could this be?" But some sneered and said, "They are full of new wine!" (HCS)

The reference to tongues and languages is indicative of various people groups becoming influenced by an encounter with what Scripture calls "Holy Spirit", and I strongly suspect is the influence of helium. There was no Periodic Table of Elements back when Scripture was written and certainly none of the descriptive terms of those elements were present. Different languages (communication) is the term used to describe the varying symptoms an individual person's body could have in response to the helium encounter; and persons from all nations being described as tongues. Fire is a reference to heat and purification. The internal heat of the body can fluctuate with the incoming/outgoing of helium; heat and fire require a source of fuel. New wine is blood that has been influenced (intoxicating), causing it to be new or different.

Ephesians 5:18-19: And don't get drunk with wine, which leads to reckless actions, but be filled by the Spirit: speaking to one another in psalms, hymns, and spiritual songs, singing and making music from your heart to the Lord. (HCS) (emphasis added)

Songs and singing is an action that takes place in the blood. This interior musical is in response to the encounter with helium from the atmosphere that produces clean blood. When the blood becomes clean, removing the stench spoken of in Isaiah 3:24, it will sing, resulting in an internal dance, a movement that is harmonious and no longer disruptive. This singing may coincide with the sounds recorded by NASA that are heard in outer space. The frequencies present in the Heavens produce a musical hum. NASA's website and

Youtube have recordings of this harmonious humming sound orchestrated by frequencies. When our blood is clean and has a healthy level of Star Dust, it will have a harmonious hum that puts us in unison with the sounds of outer space. The heart provides the rhythm for the music. (Isaiah 3:24; 34:3; Ecclesiastes 10:1)

A Word of Caution

Matthew 12:30-31: Anyone who is not with Me is against Me, and anyone who does not gather with Me scatters. Because of this, I tell you, people will be forgiven every sin and blasphemy but the blasphemy against the Spirit will not be forgiven. Whoever speaks a word against the Son of Man, it will be forgiven him. But whoever speaks against the Holy Spirit, it will not be forgiven him, either in this age or in the one to come. (HCS)

Blasphemy comes not only in the form of speech but also by action. Forgiven is descriptive of the removal of contaminations in the blood that result in damage to DNA. When you are not participating in the required steps for the True Sabbath Meditation, you are committing blasphemy. Ouch!

CHAPTER XVIII
RIVER OF LIGHT

Exodus 7:18, 21: The fish in the Nile will die, and the river will stink; the Egyptians will not be able to drink its water. Verse 21: The fish in the Nile died, and the river smelled so bad that the Egyptians could not drink its water. Blood was everywhere in Egypt. (NIV)

I bring this Scripture into play to direct the attention to the Nile, the smell, the water and the reference to blood. Nile means: River of Light. This River of Light is the plasma in the blood when combined with the helium that emits a glow. The life (or light) within the blood has died and the entire physical body has a stench to it. Body odor is a result of contaminated blood, damaged DNA. Some translations use the words "river" or "stream" in place of Nile.

Egypt is a symbol of worldly systems that have enslaved us; worn us down and caused pain and suffering. These worldly systems have caused the River of Light within the blood to die out. To bring this River of Light back to life, a weekly Sabbath Meditation practice must take place along with proper eating habits. If you haven't already guessed, this Sabbath Meditation has absolutely nothing to do with gathering in crowds on Saturday or Sunday for any type of event, including religious gatherings. To shed further light on the subject, being in a crowd on a Sabbath Meditation weekend has had its share of contributing to the River of Light being extinguished. This River of Light is what nourishes the Soul.

CHAPTER XIX
THE FINAL ACT

The final proof of what has been set forth in this book is found in I John 5. With the Spirit of Revelation to assist, the Scripture sets forth the full description of what the name "Jesus" is:

I John 5:6-9: Jesus Christ – he is the One who came by water and blood, not by water only, but by water and by blood. And the Spirit is the One who testifies, because the Spirit is the truth. For there are three that testify: the Spirit, the water, and the blood – and these three are in agreement. If we accept the testimony of men, God's testimony is greater, because it is God's testimony that He has given about His Son. The one who believes in the Son of God has this testimony within him. (HCS)

The name "Jesus Christ" stands for: the helium (Spirit), plasma (water), blood combination that brings an influence and results in evidence of salvation (rescue). Rescue from what? From contaminated blood and damaged DNA/cells that produce disease. Christ is a reference to the status in the blood; Jesus is the Sabbath Meditation (quieting yourself; being alone) that draws the Spirit (helium), that joins to the water (plasma) and moves with the blood. We could name the Sabbath Meditation influence and ultimate results thereof, Jesus. The name Jesus means save or salvation; salvation means rescue. Son references a byproduct of God; an heir, one who receives the benefit. Are you a Son?

When Scripture speaks of "the One", the combined helium, water, blood activity is what is being referred to. Testify or testimony is an act of giving proof; evidence; speaking the truth. Herein, the true Trinity is also set forth: helium, plasma, blood.

Romans 13:11-13: Besides this, knowing the time, it is already the hour for you to wake up from sleep, for now our salvation is nearer than when we first believed. The night is nearly over, and the daylight is near, so let us discard the deeds of darkness and put on the armor of light. Let us walk with decency, as in the daylight; not in carousing and drunkenness; not in sexual impurity and promiscuity; not in quarreling and jealousy. But put on the Lord Jesus Christ and make no plans to satisfy the fleshly desires. (HCS)

The time has come! What is set before you is not a design for an easy walk through life. The world has numerous foods, functions, travels, traditions and celebrations that all on some level eat away at the Star Dust and eliminate the Hebrew DNA. If you desire to live out the remainder of your days healthy and without debilitating disruption it is vital that the Star Dust is present in your blood. To advance on to an eternal hereafter when this physical life is an its end, optimal Star Dust is required. Each is responsible for their own salvation. Eternal life, or eternal death?

John 15:19: If you were of the world, the world would love you as its own. However, because you are not of the world, but I have chosen you out of it, the world hates you. (HCS)

CONCLUSION

THE WONDERS ABOVE

The sun, the moon and the stars above
This is where we find true Love.
They give of themselves and put on a show
All in an attempt to help us glow.
That glow we need to help us fight
The dangers we encounter while on this plight.
They twinkle and shine and dance about
To display their message with a shout.
Shifting and changing to their hearts content
I wish I knew what all this meant.
The components they share do great things
When you position yourself to receive what they bring.
The clouds escort their gifts to our door
Rest and see what is in store.
For thousands of years, it was all highly adored.

Izauh 61™

RESOURCES:

Holy Bible: Holman Christian Standard; Christian Standard Bible and New International Version.

TheFACTFile

Cdc.gov

KnowyourDNA.com

Nationalgeographic.com

Suggested Reading:

From Antichrist to I AM
Food For The Journey To I AM
published 2022, Harvest of Healing, LLC

Home-Made Answers for Cancer
And Life Altering Disease
published 2024, Harvest of Healing, LLC